ENGELBERTE EBODE BALLA FOUDA

Cytopathological profile of cervical cells in Cameroon

ENGELBERTE EBODE BALLA FOUDA

Cytopathological profile of cervical cells in Cameroon

Factors associated with the development of cervical cancer in rural and urban areas

ScienciaScripts

Imprint
Any brand names and product names mentioned in this book are subject to trademark, brand or patent protection and are trademarks or registered trademarks of their respective holders. The use of brand names, product names, common names, trade names, product descriptions etc. even without a particular marking in this work is in no way to be construed to mean that such names may be regarded as unrestricted in respect of trademark and brand protection legislation and could thus be used by anyone.

Cover image: www.ingimage.com

This book is a translation from the original published under ISBN 978-620-6-72016-4.

Publisher:
Sciencia Scripts
is a trademark of
Dodo Books Indian Ocean Ltd. and OmniScriptum S.R.L publishing group

120 High Road, East Finchley, London, N2 9ED, United Kingdom
Str. Armeneasca 28/1, office 1, Chisinau MD-2012, Republic of Moldova, Europe
Printed at: see last page
ISBN: 978-620-8-02389-8

TABLE OF CONTENTS

DEDICATION

"To my parents, PIERRE EBODE and FOUDA ENGELBERTHA

ACKNOWLEDGEMENTS

To Almighty God for helping me to write and complete this thesis.

To my supervisor, Dr KOANGA MOGTOMO Martin, for his invaluable and judicious advice. His dynamism and scientific skills have enabled me to successfully complete my studies over the last five years in the Faculty of Science.

To the FS/UD for having trained me over the last five years, and to the teachers in the Biochemistry Department for their expertise in our work up to this DEA examination.

To my classmates from the 2015-2016 year for the moments shared together.

To all the women who took part in this study and without whom this report would not have been possible.

The health institutions where our work was carried out.
To my brothers and sisters: EBODE EKANI Pierre, EBODE NGA Adéline, EBODE NTSAMA Sabine, EBODE BALLA Rameaux, EBODE Pierre, EBODEMVOGO Edouard and EBODE BIBOUGOU Rébecca. Thank you for always being there for me whenever I needed them.

To Abbé ADALA Antoine Marie for his moral and financial support for my studies over the last two years.

To my baby NKO'O Rita for her love
To all my friends and to all those who helped me in any way in the preparation of this dissertation.

SUMMARY

Cervical cancer is a real public health problem in developing countries, and in Cameroon in particular. In general, geographical data on the incidence or mortality of this cancer vary widely, sometimes to the advantage of rural communities and sometimes to their detriment. With this in mind, we conducted a pilot cross-sectional study in one rural (Niété) and one urban (Yaoundé I) area. The aim of this study was to determine the cytopathological profiles of cervical cells in the Niété and Yaoundé I districts, and to carry out a comparative analysis of these profiles by highlighting the risk factors responsible for their variation. The Papanicolaou technique used to determine cytological profiles yielded 9.8% pathological smears in Niété, including 4.6% ASCUS, 1.3% AGUS, 2.6% LSIL and 1.3% HSIL, compared with 16.3% pathological smears in Yaoundé I, including 1.3% ASCUS, 3.3% AGUS, 9.8% LSIL and 2% HSIL. Comparative analysis of these profiles using the Student, Fischer and Chi-square tests of Graph Pad prism v.5 software revealed a significant difference (P= 0.034; χ^2 = 11.01) between the cytopathological profiles of women surveyed in Niété and Yaoundé I, the urban area being more prone to precancerous lesions than the rural area. In addition, the age of the participants and alcohol consumption were determined to be responsible for the variation in the ASCUS type cytopathological profile between Niété and Yaoundé I, parity was responsible for the variation in the AGUS type, the number of sexual partners, marital status and again the age of the participants were responsible for the variation in the LSIL type, and contraception, the variation in the HSIL type. There was therefore an urgent need to organise well-targeted cervical cancer screening programmes in Cameroon, since the majority of women who tested positive for lesions had never heard of the cervico-vaginal smear (80%), and to do so using a methodology that was acceptable in terms of cost and effectiveness.

Key words : Cytopathological profile - cervical cell - precancerous lesion - cervical cancer - cervico-vaginal smear

INTRODUCTION

Cervical cancer is a large, uncontrolled proliferation of abnormal cervical cells in the cervix (Carozzi et al., 2016). In general, this cancer begins with structural and morphological changes in the cells of the endocervix and ectocervix, giving rise to the appearance of so-called precancerous lesions (Embolo et al., 2016). The appearance of precancerous lesions follows the transformation of normal epithelium into dysplastic epithelium (Maclean, 2009). Worldwide, cervical cancer is the fourth most common cancer in women and ranks 7e globally (WHO, 2015). It causes significant morbidity and mortality, with more than 500,000 new cases and over 300,000 deaths per year worldwide (Khenchouche et al., 2013), over 80% of which occur in developing countries (WHO, 2007). In Africa, it remains the leading cancer in women, with Central and East Africa recording the highest cancer-related mortality rates in the world, at 22.2 and 27.6 per 100,000 patients respectively, compared with 2 per 100,000 patients in East Asia and Western Europe (Somé et al., 2016). In Cameroon, the prevalence of precancerous cervical lesions is estimated at 80.73/100,000 women (Kemfang et al., 2015), and represents the second leading cause of cancer death in women, after breast cancer, in terms of mortality and incidence (WHO, 2014).Cervical cancer is characterised by the appearance of asymptomatic precancerous lesions that persist for many years (Moscicki et al., 2012). Most often, these lesions will regress spontaneously and only a small number will progress to invasive cancer (Diouri, 2008). The natural history of cervical cancer, the possibility of regular sampling and the availability of effective treatment for pre-invasive lesions mean that this type of cancer lends itself well to screening. However, it can be prevented in almost all cases if detected in the early, asymptomatic stages. Early detection of precancerous lesions therefore remains the mainstay of the fight against the disease. The disease should not be a death sentence, even in poor countries. Inexpensive, low-tech screening tools are available today. They could significantly reduce the burden of cervical cancer deaths in these countries (Bosch et al., 2013). Cytology (cervical smear or Pap test) is the most widely used technique for detecting precancerous lesions. It has dramatically reduced the incidence of invasive cancers and mortality in most developed countries.A number of human and technical problems can be noted with regard to the real cause of the resurgence of lesions. Firstly, there is a reluctance on the part of the women who are supposed to take the smear test. Secondly, there is a low proportion of women who have undergone gynaecological check-ups during or after their pregnancies. However, the Pap smear, the main tool for screening for lesions, suffers from low sensitivity and specificity (Embolo et al., 2016). Dysplastic lesions of the cervix remain poorly documented in rural Cameroon. In the study carried out in Bali, a rural locality in the North-West Province, by Tebeu et al (2005), it is important to organise screening and treatment campaigns for precancerous lesions in other rural and urban localities in Cameroon in order to establish a national programme for the management of these lesions for better prevention of cervical cancer deaths in Cameroon. For several years now, rural areas have been going through a major crisis. They are experiencing a massive exodus to cities and metropolises. In such a context, it is legitimate to ask whether living in a rural community rather than in a city has an impact on the health and well-being of this population. This is the background to the present study. Is there a real difference between the cytopathological profiles of cervical cells in rural and urban women in Cameroon? What might be the causes? To answer these questions, we started from the hypothesis that women living in the Arrondissements of Niété (rural area) and Yaoundé I (urban area) would have a different cervical cytopathological profile. This made it possible to set ourselves the general objective of listing the factors responsible for the variation in cell morphologies in women from Niété and Yaoundé I, and the specific objectives of :

➤ To determine the cytopathological profile of the cervical cells of women living in the Niété and Yaoundé I districts.
➤ Carry out a comparative analysis of these profiles

➤ Identify the factors involved in varying these profiles.

CHAPTER I

LITERATURE REVIEW

I.1. EPIDEMIOLOGY OF CERVICAL CANCER

I.1.1 World situation

With 528,000 new cases each year (IARC, 2013), cervical cancer is the fourth most common cancer in women worldwide, after breast, colorectal and lung cancers (Figure 1). It is also the sixth most common cause of death from cancer (Figure 2) (Siegel et al., 2016). There is, however, a marked inequality in the distribution of incidence between countries (Bray et al., 2013). Around 85% of cervical cancers occur in developing countries, where access to screening and care, high parity and the general environment are completely different from those in developed countries (Bosch et al., 2013; IARC, 2013). The regions where cervical cancer incidence and mortality are among the highest in the world are (Bray et al., 2013): sub-Saharan Africa, Latin America and South Asia (ACCP, 2004).

Cancer	Cases	%
Breast cancer	246,660	29%
Lung cancer	106,470	13%
Colorectal cancer	63,670	8%
Cervical cancer	60,050	7%
Thyroid cancer	49,350	6%
Lymphoma	32,410	4%
Skin cancer	29,510	3%
Leukaemia	26,050	3%
Pancreatic cancer	25,400	3%
Kidney cancer	23,050	3%
Other	843,820	100%

Figure 1: Estimated incidence of the top ten cancers (Siegel et al., 2016)

Lung cancer	72,160	26%
Breast cancer	40,450	14%
Colorectal cancer	23,170	8%
Pancreatic cancer	20,330	7%
Ovarian cancer	14,240	5%
Cervical cancer	10,470	4%
Leukaemia	10,270	4%
Liver cancer	8,890	3%
Lymphoma	8,630	3%
Brain cancer	6,610	2%
Other	**281,400**	**100%**

Figure 2: Estimated mortality rate for the top ten cancers (Siegel et al., 2016)

I.1.2 Situation in Africa

Cervical cancer is a preventable disease. However, in Africa it is the leading cancer in women, with Central and East Africa recording the highest mortality rates from this cancer in the world, at 22.2 and 27.6 per 100,000 patients respectively, compared with 2 per 100,000 patients in East Asia and Western Europe (Somé et al., 2016).

I.1.3 Situation in Cameroon

The prevalence of precancerous cervical lesions is estimated at 7% in Cameroon (Criton, 2010). In 2002, these precancerous lesions were the leading cause of death from cancer in women with an average age of 49, accounting for 32% of all cancers and 12% of new cases discovered each year out of 10,000 (ROBYR, 2002). Following the adoption of screening programmes using FCU and raising public awareness of the need for screening, it has been possible to reduce the mortality rate of this cancer considerably. It is currently the second leading cause of cancer death in women, after breast cancer, in terms of mortality (Figure 3) and incidence (Figure 4) (WHO, 2014).

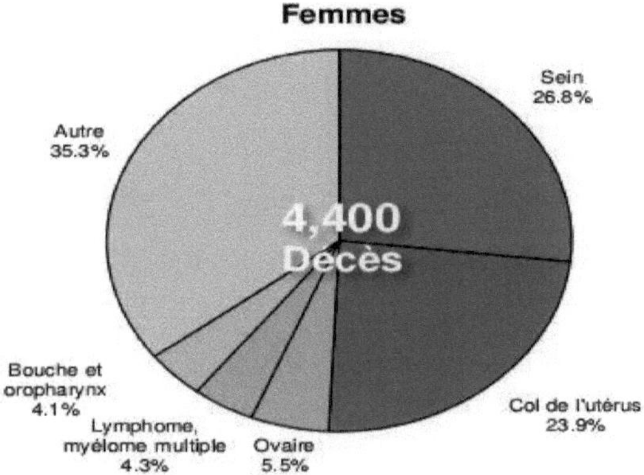

Figure 3: Cancer mortality profile in Cameroon (WHO, 2014)

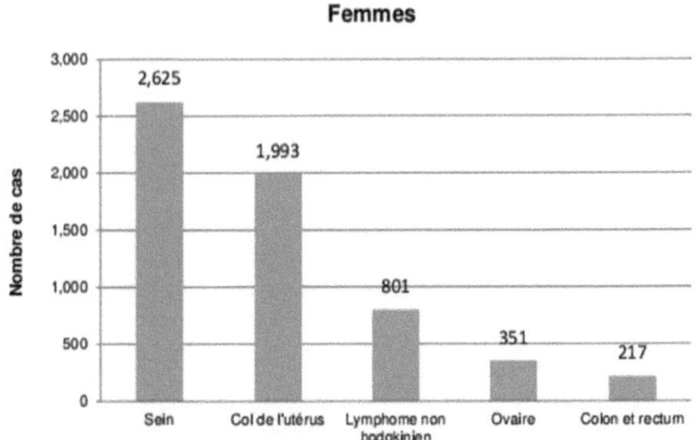

Figure 4: Cancer incidence in Cameroon (WHO, 2014)

I.2 GEOGRAPHICAL SITUATION OF THE DISTRICTS OF NIETE AND YAOUNDE I

The Borough of Niété has a population of 40,894, divided between 28 villages, covering an area of 2,117 km^2 . It is located in the South Region, Ocean Department, bounded by :

✓ To the south by the municipality of Campo,

✓ To the north by the Commune of LOKOUNDJE,

✓ To the east by the Commune of AKOM II,

✓ To the west by the Communes of Lokoundje and Kribi I.

The Borough of Yaoundé I is located in the Department of Mfoundi, Centre Region. It has a population of 281,586 over an area of 55 km^2 . The Borough of Yaoundé I is bounded by

✓ To the north by the OKOLA Rural Commune (LENDOM village),

✓ To the south by the Commune d'Arrondissement de Yaoundé V (EKOE-rails stream),

✓ In the south-west, via the Yaoundé III district, in particular via the Mfoundi river and the Boulevard du 20 MAI.
✓ To the west, via the Yaoundé II District Council (Warda crossroads, new Bastos road),
✓ To the east and north-east by the Commune d'Arrondissement of SOA

I.3 ANATOMY OF THE CERVIX

The uterus is a smooth, pear-shaped hollow muscle with thick walls (Figure 5).

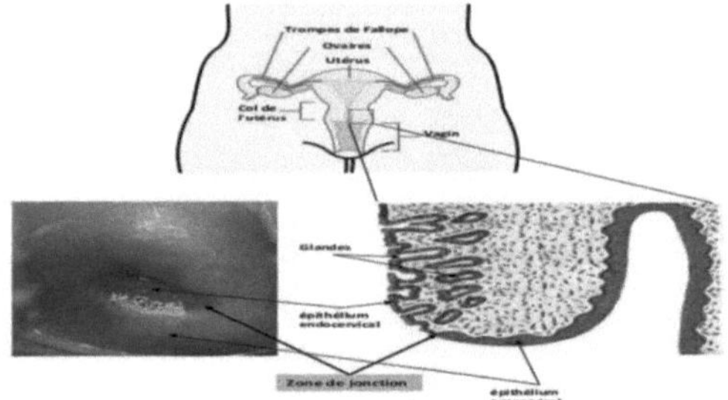

Figure 5: Anatomical and histological structure of the cervix (Chard, 1994; James, 2016)

The cervix is lined by two types of epithelium: non-keratinised squamous epithelium (or malpighnian epithelium) and columnar epithelium (or glandular epithelium). (Dallenbach, 2006)

> **Exocol:**

The stratified squamous epithelium is made up of several layers o f increasingly flat cells (Sellors, 2004). Opaque and pale pink in colour before the menopause, this epithelium is made up of four layers of cells:

✓ The lower basal layer (CB)

✓ The parabasal layer (CPB)

✓ The intermediate layer (CI)

✓ The surface layer (SL)

After the menopause, the squamous epithelium thins, takes on a whitish-pink colour and becomes more fragile, so more sensitive to trauma, which often results in small haemorrhages or petechiae (Sellors, 2004).

➢ **The endocol:**

The columnar epithelium consists of a single layer of high cells resting on the basement membrane. This layer is made up of mucus cells and ciliated cells, with subcylindrical, bipotential cells known as "reserve cells" interspersed between them. This epithelium invaginates into the underlying connective tissue to form muco-secreting crypts, the orifices of which open at the level of the surface coating (WHO, 2007).This epithelium is therefore much thinner than the squamous epithelium lining the ectocervix. On examination with the endocervical speculum, it appears bright red.

➢ **The original squamocolumnar junction (SCJ)**

It appears as a narrow line, marked by a dip due to the difference in thickness between the squamous and columnar epithelia. The location of the original SJC varies with a woman's age, hormonal status, the trauma of childbirth and whether or not she uses oral contraception (WOLFGANG, 2003). When exposed to vaginal acidity, the columnar epithelium is progressively replaced by stratified squamous epithelium, consisting of a basal layer of polygonal cells derived from subepithelial reserve cells. This normal physiological process is called squamous metaplasia and gives rise to a new squamous epithelium. Once it has matured, the neoformed squamous epithelium closely resembles the original squamous epithelium. However, on visual examination, the neoformed squamous epithelium is different from the original squamous epithelium. The remodelling zone corresponds to the area where squamous metaplasia has occurred, where the columnar epithelium is, or has been, replaced by squamous epithelium.

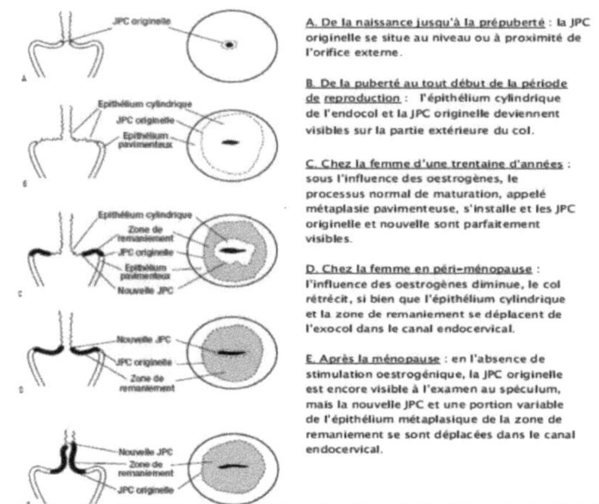

Figure 6: Process of squamous metaplasia: (WHO, 2007).

I.4 NATURAL HISTORY OF CERVICAL CANCER

The stratified squamous epithelium lining the cervix provides protection against toxic substances and infection. Under normal conditions, the upper layers are constantly renewed, ensuring that the integrity of the epithelial lining is maintained through the constant and orderly formation of new cells in the basal layer.Sexual intercourse is the classic route of contamination by HPV viruses, but these naked viruses, which are highly resistant to temperature, can be transmitted by vectors such as water, linen, equipment and soiled gloves. HPV penetrates the basal cells of the epithelium, probably at the ectocervical junction, which is particularly vulnerable to all kinds of aggression. (Letian and Tianyu, 2010 ; Trimble et al., 2010) Epithelial infestation by the virus generally takes place through a micro-lesion and requires contact and penetration of the viral particles into the basal epithelial cells. Contact is facilitated by cofactors:

➢ recurrent cervicitis: this can lead to cervical microlesions and/or potentiate HPV-related cytopathogenic lesions (trichomonas vaginalis cervicitis).
➢ Ectropion: the area outside the junction can be easily damaged.

➢ polyps: they are fragile and present a risk of cancerisation
The cytopathogenic effect of HPV replication results in the presence of koilocytes: vacuolated cells with large nuclei, the presence of which is

pathognomonic of HPV infection. Cervical cancer develops very slowly, over a period of around fifteen years. The precursor lesions generally begin at the junction between the squamous and glandular mucosa of the cervix. It is therefore important to locate this constantly changing area, as it is a prime site for HPV infection. HPV infects the basal cells of the squamous epithelium, usually through microtrauma. Cervical cells infected in this way take on a particular appearance: the nucleus is surrounded by a pale halo with irregular contours, corresponding to an area of cytoplasmic necrosis. These are known as koilocytes (Figure 7). This cytopathic effect is specific to HPV infection (Baba et al., 2007).

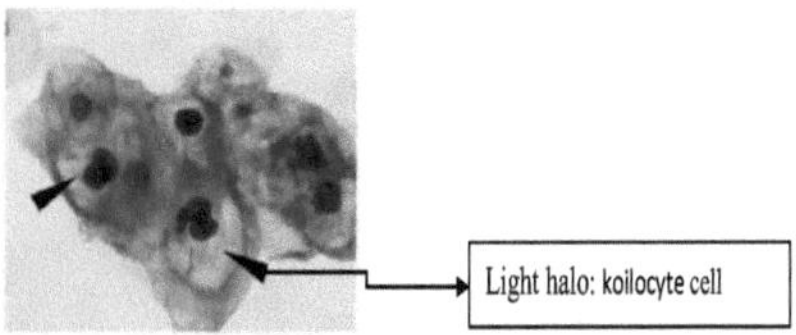

Figure 7: Appearance of a koilocyte: cell whose nucleus is surrounded by a light-coloured halo, cytopathic effect of HPV infection (Chard, 1994).

Naturally, cervical cancer is preceded by precancerous lesions that persist for years before progressing to a low-grade lesion on cytology, which is broadly a linear pattern: mild (CIN1), moderate (CIN2) and severe (CIN3) dysplasia, carcinoma in situ, microinvasive carcinoma and finally invasive carcinoma. (Lecuru, 2008)Histologically, progression results in the loss of cellular differentiation, giving the appearance of cervical intraepithelial neoplasia (CIN). This progresses from CIN 1 to CIN 3, then to invasive cancer. Between 10% and 15% of untreated CIN 1 will progress to CIN 2-3, while the remainder will regress spontaneously within two years of initial diagnosis. The risk of progression from CIN 1 to CIN 3 or a more severe lesion has been estimated at 1% per year, while the risk of progression from CIN 2 to the more severe stages would be from 16% in 2 years to 25% in 5 years (Moscicki et al., 2012).

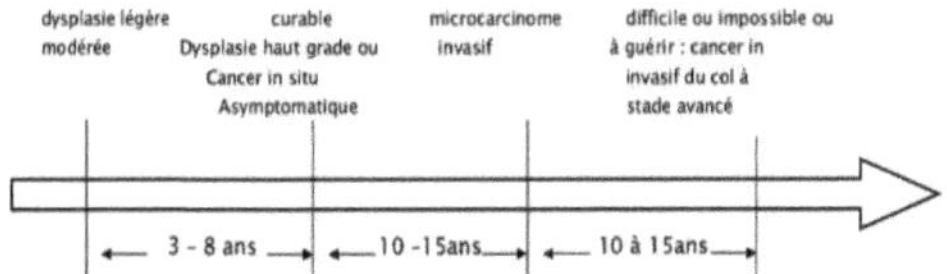

Figure 8: Sequence of events in the natural history of cervical cancer (Moscicki et al., 2012).

For each precancerous cervical lesion, there is a probability of regression (from 32% to 57% depending on the severity of the lesion) towards a normal epithelium, accompanying viral clearance, and a probability of persistence or progression to a more advanced stage, including for carcinomas in situ assimilated to CIN 3 (McCredie et al., 2008).

I.5 RISK FACTORS

Epidemiological data show that cervical cancer is a multifactorial disease, with HPV appearing to be the most important factor in the genesis of cervical cancer, but other co-factors are also necessary. (Lecuru, 2008) The incidence of invasive cervical cancer has fallen thanks to widespread screening in developed countries (P. Lopes, 2013).

I.5.1 Infectious factors:

➢ HPV infection:

The vast majority of cancers are caused by HPV, a sexually transmitted agent that infects the cells of the cervix,   causing cellular changes that can progress to cancer (Monsonégo, 2006). The relative risk of the association between HPV and cervical cancer is two to three times higher, compared with other cancer risk factors. (Koutsky et al., 1992; Faysal et al., 2016)

➢ Chlamydia trachomatis:

It is a very common sexually transmitted agent, causing cervicitis and metaplasia which facilitates HPV infection (Golijow et al., 2005).

➢ Herpes Simplex2 virus:

Although attention has gradually focused on HPV, it appears that herpes simplex virus plays a role in the genesis of dysplastic lesions (Boulanger, 2010). The combination of HPV and type 2 herpes increases the risk of cervical cancer by a factor of two to three (Smith et al., 2002).

> **Other sexually transmitted diseases:**

Cytomegalovirus (CMV) and human immunodeficiency virus (HIV) are considered to be cofactors that increase the risk of developing precancerous lesions, but this moderate increase in risk is controversial, depending on the study (Schiffman et al., 1995; Mougin et al., 2001).

I.5.2 Sexual and obstetric factors:

The sexual and obstetric factors often found are: (Leidy, 1999)
✓ Early sexual relations.

✓ Early marriage.

✓ Young age at first pregnancy.

✓ Multiple pregnancies.

✓ Multiple sexual partners.
Several factors increase the risk of cervical cancer, the most convincing and consistent being multiple sexual partners and early sexual activity. Early sexual activity is thought to be an important risk factor because during puberty, cervical tissue undergoes various changes that can make this area more vulnerable to lesions. (Leidy, 1999; OUCHEN, 2009)High parity is a factor linked to trauma during childbirth, but also to hormonal and immunological changes during pregnancy which favour metaplasia and the development of HPV (Leidy, 1999). High parity is linked to the progression of HPV infection, due to cervical ectropion, which is more prevalent in multiparous women and increases with the number of pregnancies (INC, 2013).

I.5.3 Smoking ;

Active and passive smoking are significantly associated with cervical lesions. Smokers have twice the risk of cervical cancer. In fact, the number of cigarettes smoked per day is correlated with the severity of the disease, and women who smoke more than 10 cigarettes per day have a greater risk of high-grade cervical intraepithelial lesions. (Leidy, 1999; INC, 2013) Smoking appears to prevent the spontaneous healing of precancerous lesions, thereby allowing the progression to cancer. It also reduces the immune response, increasing the risk of persistent infection (INC, 2013). The effect of tobacco on the epithelium of the uterine cervix can be explained by the fact that derivatives of these compounds are distributed in bodily fluids, and the nicotine is even concentrated in cervical mucus. Another constituent of smoke: cotinine (a chemical which is made by the body from the nicotine in cigarette smoke) is also found in this mucus, even in some non-smoking women who are victims of passive smoking (INC, 2013).

I.5.4 Oral contraception :

Oral contraception is considered to be a potential factor in the development of cervical cancer. (Leidy, 1999)The choice of contraceptive method also appears to affect the risk of cervical cancer, with barrier methods appearing to reduce the risk, while oral contraceptives appear to increase it (Mantovani et al., 2001).

I.5.5- Socio-economic status :

It is universally recognised that invasive cervical cancer affects women from low socio-economic backgrounds in particular, due to their limited income, poor hygiene and lack of preventive behaviour. (Muñoz et al., 2006; INC, 2013)

I.5.6 Immune status :

Immunosuppressed women are at greater risk of developing dysplastic lesions, as the risk of viral infection is 17 times higher than in the rest of the population. As a result, regular cytological surveillance is essential, particularly in HIV-positive patients, kidney transplant recipients and dialysis patients. (Leidy, 1999)

I.5.7 Nutritional factors:

Consumption of alcoholic beverages, cruciferous vegetables (cabbage, turnips, broccoli) and saturated fats increases the risk of cervical cancer. (Leidy, 1999; INC, 2013; Potischman et al., 1996).

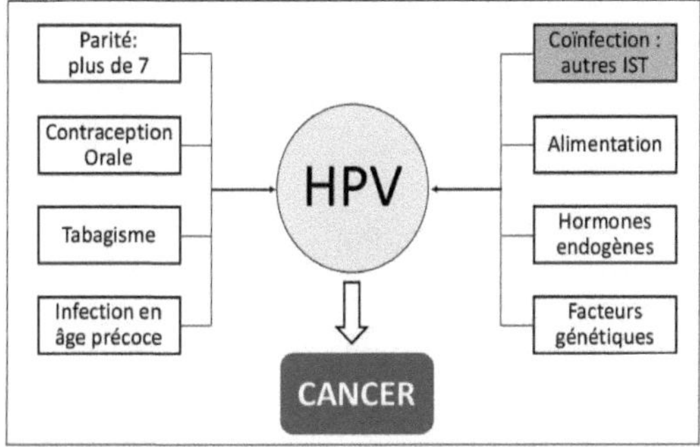

Figure 9: Risk factors for cervical cancer (Luna et al., 2013)

I.6 CERVICAL CANCER SCREENING :

Screening is a public health action carried out on a population at risk, known as the target population; it consists of detecting women with precancerous lesions and treating them (OUCHEN, 2009). It was defined in 1957 by the USCCI (the United States Commission on Chronic Illness) as "the presumptive identification of a lesion or disorder not known to the subject, by applying tests or examinations or other procedures that can be applied rapidly" (OUEDRAOGO, 2015). Thus the aim of screening is not to diagnose a disease, but to identify individuals who have a high probability o f contracting or developing it.

I.6.1 Cervical cancer screening tests and methods

Reference screening is a test that is accurate, reproducible, inexpensive, easy to perform and interpret, acceptable and safe. A multitude of procedures meet this description to varying degrees:

➢ Cytology: conventional or liquid-based

➢ Test for human papillomavirus DNA

➢ Visual inspection w i t h acetic acid (VIA) or Lugol's iodine solution (VILI).

I.6.2 Cytology or cervico-vaginal smear:

➢ Age at which smears were taken:

The first smear should be taken within a few months of sexual debut. The authors have placed particular emphasis on the 25 to 30 age group, because dysplasia has a peak incidence rate in this age group, and in situ cancer peaks in the 31 to 35 age group, 15 years before invasive cancer peaks. (Ronco, 2009)

➢ Frequency of smear tests:

A smear test every ten years reduces the risk by almost two-thirds at a much lower cost, and can therefore be a good screening strategy in countries where resources are limited. For example, it is better to screen the entire population every ten years than half the population every five years or 30% every three years, the latter strategy leading to twice as many women being screened (SAWAYA, 2003; Benchimol, 2014; Sawaya et al., 2003).

Table 1: Reduction in the rate of cervical cancer between the ages of 35 and 64 according to smear frequency, after two negative smears. (Miller, 2012)

Frequency of smear	1 year	2 years	3 years	5 years	10 years
% reduction rate cumulative	93,3	93,3	91,4	83,9	64,2

I.7 CYTOPATHOLOGICAL CLASSIFICATIONS

Pre-cancerous lesions are referred to as either dysplasia (mild, moderate or severe) according to the World Health Organisation (WHO) classification, or cervical intraepithelial neoplasia according to the Richart classification. In the end, however, the Bethesda classification is used as the reference according to the recommendations of the Agence Nationale d'Accréditation et d'Evaluation en Santé (ANAES), following the abandonment of the Papanicolaou classification (Solomon et al., 2002).

I.7.1 The BETHESDA classification

The Bethesda system was first used in the USA in 1988, and was re-evaluated in 1998 and more recently in 2001. Currently, the interpretation of FCV by pathologists is based on the Bethesda 2001 system. (Solomon et al., 2002) This classification allows correspondence between cytology and histology. Cervical and vaginal lesions are classified as squamous intraepithelial lesions (SIL), which are subdivided into low-grade and high-grade lesions.

Under this system, a smear report must consist of three parts:
✓ The first part demonstrates the interpretability of the smear.

✓ The second part reports any abnormalities of the squamous and/or glandular cells.
✓ The third part provides recommendations and clarifications.

➤ **Squamous or epidermoid cell abnormalities:** (Solomon et al., 2002)
- Epithelial cell atypia (ASC): Of undetermined significance (ASC-US) or High grade squamous intraepithelial lesion cannot be excluded (ASC-H)

- Low-grade squamous intraepithelial lesions (LSIL)
- High-grade squamous intraepithelial lesions (HSIL)
- Squamous cell carcinoma

➤ **Glandular cell abnormalities:** (Solomon et al., 2002)
- Glandular cell atypia (GCA): endocervical, endometrial or no other indication.

- Glandular cell atypia in favour of neoplasia: endocervical or without other indication

- Endocervical adenocarcinoma in situ (AIS).
- Adenocarcinoma.

I.7.2 Classification of CIN

Richart introduced this classification in the 1960s, and it is still used in some countries. It classifies abnormalities into 3 stages: CIN I, CIN II and CIN III. This classification is based on knowledge of the natural history of the cervix. (CNGOF, 2016)

I.7.3 Papanicolaou classification

Papanicolaou classified smear results into 5 categories which are a source of multiple ambiguities. This classification, which dates from 1943, was abandoned by pathologists at the BETHESDA meeting in December 1988 (except in certain countries where modified forms are still used). (Renolleau, 1996)

I. = Normal epithelial cells.
II. =Epithelial cells with inflammatory cells.
III =Doubtful abnormal cells=dysplasia.
Very suspicious with small numbers of tumour cells.
V. =Many neoplastic cells=malignant tumour.
The disadvantage of this classification is that it does not distinguish between classes II and III.

I.7.4 The WHO classification.

In the 1970s, the WHO classified cytology data into 6 categories. This classification, unlike Papanicolaou's, correlates cytology and histology. (CNGOF, 2016)

Table 2: Comparison of the different classifications used to interpret cytology (Solomon et al., 2002).

System of Papanicolaou	WHO system	CIN system RICHART	Bethesda system
Class I			The limits of normal
Class II			Changes benign of cells (AUC)
Class III	Minimal dysplasia	CIN I CIN II CIN III	Low-grade squamous cell atypia (LSIL)
	Moderate dysplasia		
	Severe dysplasia		
			Squamous cell atypia of high grade (HSIL)
Class IV	Carcinoma in situ	CIN III	
Class V	Carcinoma microinvasive	Invasive carcinoma	Invasive carcinoma
	Invasive carcinoma		

I.8 CERVICO_VAGINAL SMEAR TECHNIQUES

There are two cervical-vaginal frottage techniques:
➢ Conventional FCV.

➢ Thin layer FCV or liquid suspension smear.

I.8.1 Conventional FCV

A procedure described in 1943 by Papanicolaou, the conventional cervico-vaginal smear involves taking cells from the transformation zone of the cervix, as this is where almost all high-grade lesions develop. (WHO, 2007b; Thin et al., 1975)The WHO raises the problem of how to carry out a smear test, and states that in order to carry out a smear test correctly, it is necessary to be able to observe the cervix directly at Using a speculum, an endocervical cytobrush is introduced and cells are scraped by gently rotating it 180° along the junction of the columnar and malpighnian epithelia. A second sample is taken from the endocervix using Ayre's spatula (WHO, 2007b). The cells are then spread out on a glass slide and immediately fixed to preserve their morphological state. The labelled slide is then sent to the cytology laboratory for staining, before being examined under the microscope to determine whether the cells are normal and to classify them appropriately.

I.8.2 FCV in a thin layer or liquid suspension

This test was introduced in the mid-1990s and is mainly used in environments with high resources. Cytology in a liquid medium consists of removing the cells with a "cerebrox brush" and then collecting the entire brush in a vial containing a special preservative. The sample is then sent to the laboratory, which prepares the smear. This method requires specific equipment: a shaker, a reading device specially designed for this technique and a centrifuge with adaptable slides.

There are three techniques:
✓ Thinprep cytic[R] fully automated, the most expensive,

✓ The cytorich[R] , a semi-automatic or manual autocyte,

✓ Cyleasy[R] the least expensive.

CHAPTER II
MATERIALS AND METHODS

II.1 ETHICAL CONSIDERATIONS

Ethical clearance was obtained from the institutional ethics committee of the University of Douala under number CEI-Udo/707/11/2016/T (Appendix 3).

II.2 TYPE OF STUDY

This was a prospective, descriptive, cross-sectional study. Data were collected from subjects at the time of cancer screening using a standardised questionnaire comprising four parts: social and anthropological information; gynaecological and obstetric data; information on pathology and serological status; and toxicological aspects (Appendix 4).

II.3 STUDY LOCATIONS

Samples were collected first in the Borough of Niété, where no data on cervical cancer had yet been compiled, and then subsequently in the Borough of Yaoundé I. Recent studies reported here by various authors (Sando et al., 2014; Kemfang et al., 2015) on UCC have shown that women in Yaoundé are mainly exposed to this disease, which has moreover been diagnosed as the first cancer affecting women in Yaoundé (Sando et al., 2014).

II.4 STUDY PERIOD

This study was carried out over a period of four months, from June to October 2016.

II.5 STUDY POPULATION

➢ Inclusion criteria

All sexually active women were included in the study on a voluntary basis and signed an informed consent form.

➢ Non-inclusion criteria

Any woman who was pregnant or had undergone a total hysterectomy was not included in the study. Any woman with a current menstrual period or within 5 days of menarche. menstruating, women who have had sexual intercourse 48 hours prior to sampling and women who have not given informed consent.

II.6 SAMPLING

➢ Type of sampling

Participants were recruited consecutively.

➢ Sample size

The study was based on a sample of 306 women, including 153 in the Niété district and 153 in the Yaoundé I district, determined by convenience.

II.7 MATERIALS AND METHODS

II.7.1 Pre-analytical phase

It consisted of a :

✓ Raising awareness in the field :

Raising awareness among the target population was done through radio announcements and information leaflets in the various public and private health facilities, secondary schools and training schools.

✓ Recruiting women

➢ A Niété

Contacts were made with the local authorities, the village chiefs and the doctors in charge of the health centres in the various villages (Zingui, village 13 and Adjap, its administrative centre). They were responsible for passing on information to the local population two weeks before our arrival. The village chiefs announced the campaign at their local meetings. Messages were broadcast on the radio and in churches, and leaflets were distributed in the various villages. Health centre staff also invited women attending consultations to take part in the campaign. Women were recruited on Saturday and Sunday, from 8am to 3pm.

➤ In Yaoundé I

It was easier to recruit women here, as there were quite a few women coming in each day, and the CHU nursing staff invited them to go to the room reserved for the screening campaign. Women were recruited from Monday to Friday, 8am-3pm.

✓ Survey in the form of questionnaires:

Women who had given their informed consent were asked to complete the questionnaire.

✓ Cervico-uterine sampling :

Most of the patients were undergoing this examination for the first time. After explaining the purpose of the examination, the woman was placed in the gynaecological position on the examination table, followed by the gynaecological examination. After a visual inspection of the cervix for macroscopic lesions, a speculum was inserted up to the cervix after wearing gloves. Once the cervix was accessible and inspected, the sample was taken by first scraping the cells from the vagina (using the sharp end of Ayre's spatula). Samples were then taken from the ectocervix (using the notched end of Ayre's spatula) and the endocervix (using a cytobrush). The samples were taken by a doctor from the health centre, and the slides were read by an experienced cytopathologist.

✓ Labelling :

These samples were spread on three identifiable slides, each bearing the patient's number and the area sampled. Each sample was followed by immediate fixation with Cytospray®, in order to keep them in a suitable state of hydration.

II.7.2 Analytical phase

II.7.2.1 Procedure for determining the cytopathological profile of the cervix in Niété and Yaoundé I.

Papanicolaou staining or Pap-test

> **Purpose of colouring**

Papanicolaou staining is designed to highlight squamous and endocervical cells taken from the vaginal wall and cervix.

> **Principle**

This method stains fixed cell material. It uses a polychromatic action resulting from the intervention of a nuclear dye in the aqueous phase (Harris Haematoxylin) and two cytoplasmic dyes in the alcoholic phase (Orange G6 and EA 50).

> **Technical** (Healthline, 2012) See Appendix 1
> **Comments**

Cores	Blue to blue-violet
cytoplasm of deep cells	Green to greenish blue
Cytoplasm of superficial cells	Pink or orange-red

The nuclei are stained blue-violet for superficial cells, pink-orange for intermediate cells and blue-green for cells in deeper layers. The cytoplasm is coloured pink or orange-red when eosinophilic and green or greenish-blue when basophilic.

II.7.3 Post-Analytical phase

✓ Quality control of results

✓ Checking results

✓ Results and counselling for people with abnormal smears.

II.8 STATISTICAL ANALYSIS

Statistical analyses were carried out using :

➢ Excell (2007) for managing tables and figures

➢ GraphPad Prism 5 software was used to analyse the data. Student's t-test on unpaired series was used to compare means, Fisher's F-test to compare variances and Pearson's Chi-square test of independence to compare percentages. The significance level was set at $P \leq 0.05$.

CHAPTER III
RESULTS AND DISCUSSION

III.1 PRESENTATION OF THE GENERAL CHARACTERISTICS OF THE POPULATION STUDIED

The general characteristics of the 306 participants who had a Pap smear that was deemed interpretable are shown in Tables 3 and 4. The mean age of the women was 36 ± 11 years in rural areas and 36 ± 10 years in urban areas; mean parity was 3.7 ± 2.3 in rural areas and 4.1 ± 2.5 in urban areas. The mean age at first pregnancy was 18 ± 4.8 years in Niété and 18 ± 3.3 years in Yaoundé I. More than two thirds of the participants (71.2%) were married at the time of the survey. This was the case in both study areas. The average age at first sexual intercourse was 16 ± 2.3 years in rural areas compared with 17 ± 3.1 years in urban areas, with an average cumulative number o f sexual partners of 4.1 ± 3.6 in rural areas compared with 2.2 ± 2 in urban areas.

III.1.1 Socio-demographic and educational data on women

In the study population, 40.2% (123/306) of the women surveyed were in the [25-34] age group. The cumulative frequency in urban areas for women under 45 was 75.8% and 77.8% in rural areas. Women in the 30-39 and 50-66 age groups were most represented in Niété. Women in Yaoundé I were most represented in the 17-29 and 40-49 age groups. There was no significant difference (P=0.793) in terms of age between arrondissements. More than two-thirds of the women (71.2%) were married at the time of the survey. This was the case in both areas. There was a significant difference (P<0.0001) in the distribution of marital status between the two districts. Women with higher levels of education were more represented in the urban area, while women with primary and secondary education were more represented in the rural area. There was also a significant difference (P<0.0001) in the distribution of educational levels between the two districts. Unemployed women were more represented in rural than in urban areas. There was also a significant difference between the two districts in the distribution of occupational status (P=0.0057).

Table 3: Socio-demographic and educational characteristics of patients.

Features	Rural areas		Urban environment		Total		p-value
	N	%	N	%	N	%	
Age groups (years)							
17-29	46	30,1	48	31,4	94	30,7	
30-39	52	34	45	29,4	97	31,7	
40-49	36	23,5	42	27,4	78	25,5	0,793
50-66	19	12,4	18	11,8	37	12,1	
Professional situation							
Yes	55	35,95	79	51,6	134	43,8	
No	98	64,05	74	48,4	172	56,2	0,0057*
Marital status							
Brides	111	72,5	107	69,9	218	71,2	
Single	39	25,5	29	19,0	68	22,2	
Divorced	0	0,0	4	2,6	4	1,3	<0,0001*
Widows	3	2,0	13	8,5	16	5,2	
Level of study							
No	12	7,8	17	11,1	29	9,5	
Primary	74	48,4	38	24,8	112	36,6	
Secondary	65	42,5	62	40,5	127	41,5	<0,0001*
Superior	2	1,3	36	23,5	38	12,4	

Data are presented as headcount and percentage. P-value calculated from the ordered analysis of variances (significant P-value < 0.05). (*) = significant P value (< 0.05)

III.1.2 Gynaecological-obstetric and toxicological data on women

Women were more multiparous (number of children ≥3) and pauciparous (number of children = 1 or 2) in rural and urban areas. First sexual intercourse between the ages of 15 and 19 was higher in both study areas. Rural areas were more likely to have more than three sexual partners (57.4%) and urban areas more likely to have only one partner (57.5%). Significant differences between Niété and Yaoundé I were observed for contraception (p<0.0001), age at first sexual intercourse (p=0.0193), number of sexual partners (p<0.0001), smoking (p<0.0001) and alcohol (p=0.0002).

Table 4: Gynaecological-obstetric and toxicological characteristics of patients.

Features	Rural areas		Urban areas		Total		p-value
Parity nulliparous	11	7,2	12	7,8	23	7.5	
Paucipare	33	21,7	32	20,9	65	21.2	0,984
Multipare	109	71,3	109	71,3	218	71,3	
Contraception Yes	67	43,8	126	82,4	193	63,1	
No	86	56,2	27	17,6	113	36,9	<0,0001*
Age of first sexual intercourse< 15	25	16,3	25	16,3	50	16,3	
15 à 19	113	73,8	96	62,8	209	68,4	
20 à 24	14	9,2	28	18,3	42	13,7	0,0193*
25 à 29	1	0,7	4	2,6	5	1,6	
Number of sexual partners 1	48	31,5	88	57,5	136	44,5	
2	17	11,1	28	18,3	45	14,7	<0,0001*
≥3	88	57,4	37	24,2	125	40,8	
Tobacco Yes	54	35,3	5	3,3	59	19,3	
No	99	64,7	148	96,7	247	80,7	<0,0001*
Alcohol Yes	90	58,8	56	36,6	146	47,7	
No	63	41,2	97	63,4	160	52,3	0,0002*

Data are presented as headcounts and percentages. P value calculated from the ordered analysis of variances (significant P value < 0.05).(*) = significant P value (< 0.05)

III.2 RESULTS ON FCV OF PATIENTS IN NIETE (RURAL AREA)

III.2.1 Macroscopic appearance of the cervix

Macroscopic examination of the cervix was performed prior to Pap smear sampling. It revealed that the majority of macroscopic abnormalities of the cervix were cervicitis or inflammation of the cervix (75.8%). In addition, none of these women had ever had a CVF prior to our campaign, and visits to a gynaecologist were relatively rare.

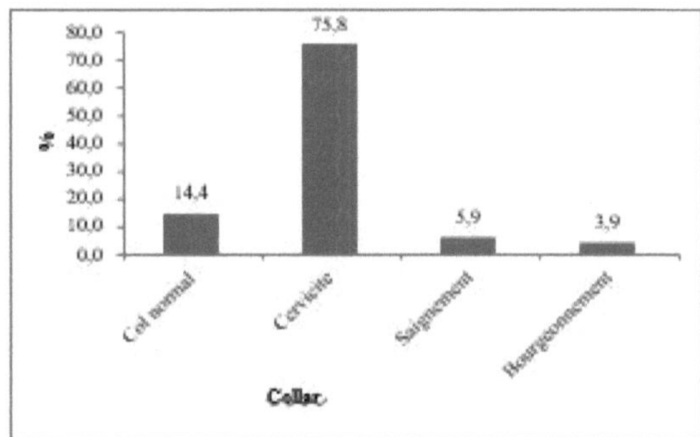

Figure 10: Macroscopic condition of the cervix of patients surveyed in Niété.

III.2.2 Cytopathological profile of cervical cells in Niété

A total of 170 patients were screened and 153 of them (90%) had smears deemed satisfactory for interpretation. The remaining 17 (i.e. 10%) with unsatisfactory smears (incomplete clinical information, poor cell spread with less than 10% of analysable cells, smears that were too inflammatory or haemorrhagic) were excluded from our study.Of the 153 smears deemed interpretable, 90.2% (138/153) were negative, and 9.8% (15/153) showed cellular atypia ranging from ASCUS to HSIL, with a predominance of LSIL-type lesions (4.6%).

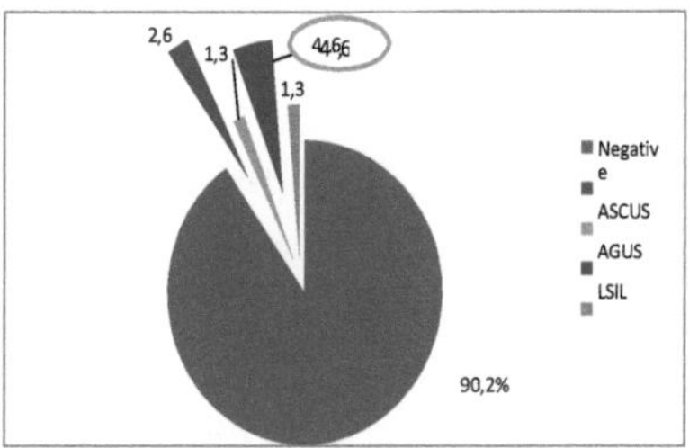

Figure 11: Cytopathological profile of cervical cells from patients surveyed in Niété.Abnormal smears= ASCUS, AGUS, LSIL and HSIL

III.3 RESULTS ON FCV OF PATIENTS IN DEYAOUNDE I (URBAN AREA)

III.3.1 Macroscopic appearance of the cervix

Macroscopic examination of the cervix in Yaoundé I revealed a normal cervix in the majority of cases (69.9%). In addition, 90% of these women had never had an FCV before our campaign, although visits to a gynaecologist were fairly frequent.

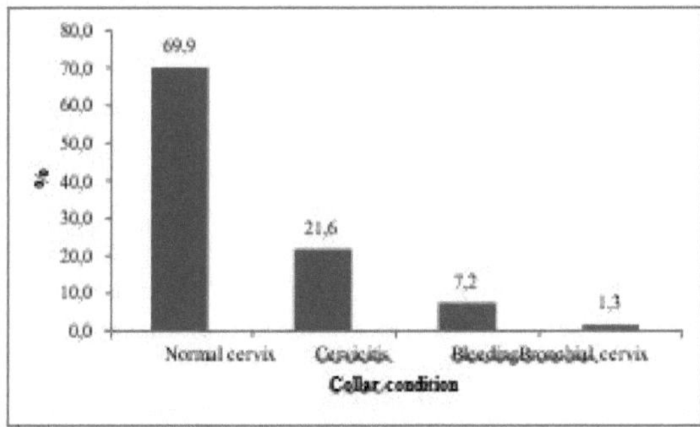

Figure 12: Macroscopic appearance of the cervix of patients surveyed in Yaoundé I.

III.3.2 Cytopathological profile of cervical cells in Yaoundé I

A total of 160 patients were screened and 153 (95.6%) had smears deemed satisfactory for interpretation. The remaining 7 (4.4%) were of unsatisfactory quality and were excluded from our study. Of the 153 smears deemed interpretable, 83.7% (128/153) were negative, and 16.3% (25/153) showed cellular atypia ranging from ASCUS to HSIL, with a predominance of ASCUS-type lesions (9.8%).

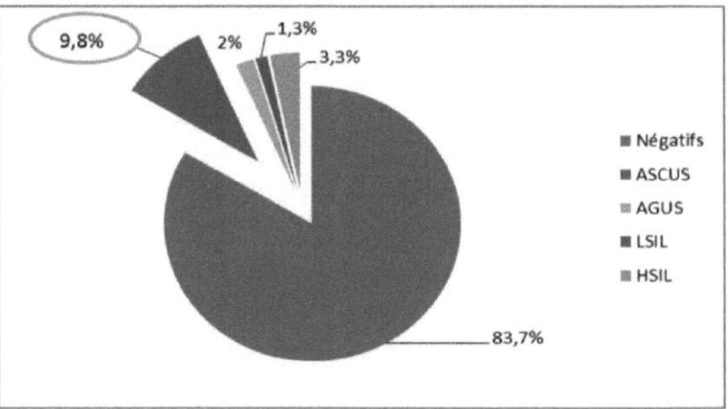

Figure 13: Cytopathological profile of cervical cells from patients surveyed in Yaoundé I Abnormal smears= ASCUS, AGUS, LSIL and HSIL

III.4 COMPARATIVE ANALYSIS OF THE CYTOPATHOLOGICAL PROFILE OF WOMEN IN NIETE AND YAOUNDE I

> **Macroscopic appearance**

A significant difference (P<0.0001; $\chi^2 = 105.6$) was observed in the distribution of the macroscopic appearance of the cervix between the two districts. Cervixes showing cervicitis or budding were more frequently encountered in Niété.

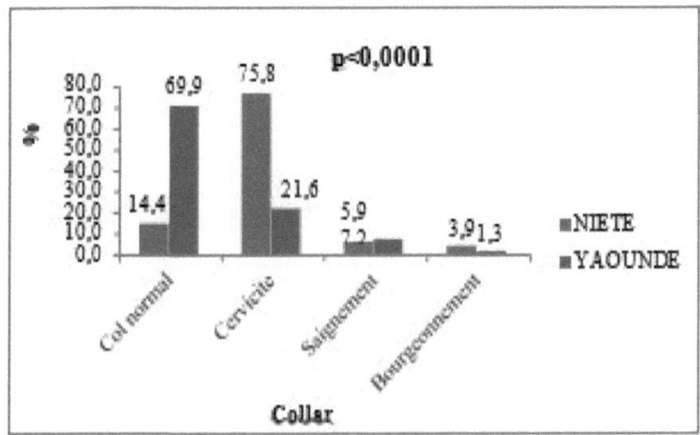

Figure 14: Comparison of cervical condition between women in Niété and those in Yaoundé I.

> **Comparison of cytopathological profiles of women's cervical cells**

The distribution of these results showed a significant difference in cytopathological profiles between Niété and Yaoundé I (P = 0.027; χ^2 = 11.01). Cytopathological profiles were more representative among women surveyed in urban areas than in rural areas (16.3% versus 9.8%).

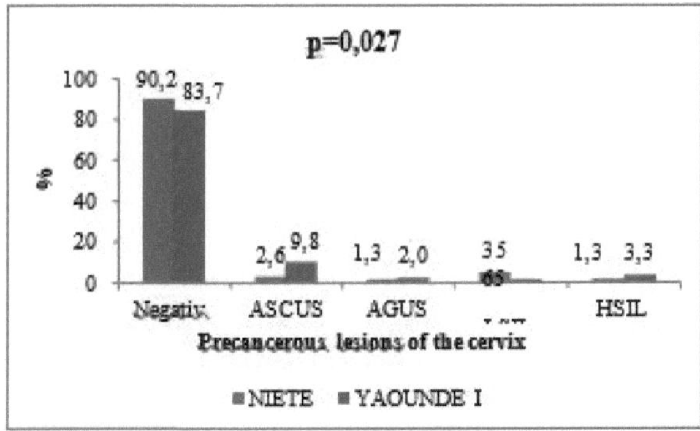

Figure 15: Comparison of cytopathological profiles of cervical cells from women in Niété and Yaoundé I

III.4.2 Factors involved in the variation in cytopathological profiles between Niété and Yaoundé I

> **Identification of factors associated with the various cytological changes observed in each Arrondissement**

Table 5 shows the effect of socio-demographic factors on the prevalence of precancerous lesions. In rural areas, it was found to be significantly associated with women aged 40 to 49 (11.1%), widows (33.3%) and women with a higher level of education (50%). In urban areas, it was found to be significantly associated with women aged 30 to 39 (19.1%), single women (48.3%) and women with secondary education (15.8%). In rural areas, the prevalence of these lesions was significantly higher among women aged between 40 and 49 (11.11%), single women (17.9%) and those with primary education (9.5%). In urban areas, on the other hand, the risk of precancerous lesions was highest among women aged between 30 and 39 (19.10%), single women (48.3%) and those with secondary education (15.8%).

Table 5: Socio-demographic factors associated with cytological changes in Niété and Yaoundé I

Socio demographics		Rural areas			Urban environment	
	N	Alteration cytological	p-value	N	Alteration cytological	p-value
Ages (years)						
17-29	46	5 (10,9%)		49	9 (18,4%)	
30-39	52	4 (7,7%)	0,004*	47	9 (19,1%)	0,005*
40-49	36	4 (11,1%)		40	5 (12,5%)	
50-66	19	2 (10,5%)		17	2 (11,8%)	
Professional situation						
Yes	52	3 (5,8%)	0,09	59	11 (18,6%)	0,08
No	101	12 (11,9%)		94	14 (14,9%)	
Marital status						
Single	39	7 (17,9%)		29	14 (48,3%)	
Brides	111	7 (6,3%)	0,042*	107	7 (6,5%)	0,009*
Divorced	0	0 (0%)		4	0 (0%)	
Widows	3	1 (33,3%)		13	4 (30,8%)	
Level of study						
No	12	1 (8,3%)		17	1 (5,9%)	
Primary	74	7 (9,5%)	0,003*	32	5 (15,6%)	0,045*
Secondary	65	6 (9,2%)		57	9 (15,8%)	
Superior	2	1		35	5 (14,3%)	

Data are presented as headcount and percentage. P-value calculated from the ordered analysis of variances. (*) = significant P value < 0.05

Table 6 shows the effect of gynaecological-obstetric and toxicological factors on the prevalence of cytological changes. In rural areas, it was significantly associated with parity (22.5%), early age at first intercourse (19.3%), high number of sexual partners (14.8%) and contraception (most often oral) (23.8%). In urban areas, it was significantly associated with early age at first intercourse (42.1%) and alcohol consumption (21.7%). However, in rural areas, the prevalence of these lesions was significantly higher in nulliparous women (18.20%), those on contraception (23.8%), those with more than three sexual partners (14.80%) and those who did not drink alcohol (25.00%). In urban areas, on the other hand, the risk of cytological changes in cervical cells was the highest. higher among women not taking contraceptives (25%) and those who drank alcohol (21.70%).

Table 6: factorsand toxicologicalassociated with cytological alterations in Niété and Yaoundé I

Gynaecological factors Obstetrics and Toxicology		Rural areas			Urban environment		
	N	Alteration cytological	p-value	N	Alteration cytological	p-value
Parity						
nulliparous	11	2 (18,2%)	0,008*	12	1 (8,3%)	0,097
Paucipare	33	5 (15,2%)		31	5 (16,1%)	
Multipare	109	8 (7,3%)		110	19 (17,3%)	
Contraception						
Yes	80	19 (23,8%)	0,003*	115	8 (7%)	0,005*
No	73	6 (8,2%)		28	7 (25%)	
Age of first sexual intercourse						
< 15	23	2 (8,7%)	0,021*	20	5 (25%)	0,039*
15 à 19	104	11 (10,6%)		82	14 (17,1%)	
20 à 24	10	1 (10,0%)		22	6 (27,3%)	
Number of partners Sexual						
1	48	2 (4,2%)	0,049*	88	13 (14,8%)	0,11
2	17	0 (0%)		28	4 (14,3%)	
≥3	88	13 (14,8%)		37	8 (21,6%)	
Tobacco						
Yes	54	5 (9,3%)	0,089	5	0 (0%)	0,48
No	97	0 (0%)		148	25 (16,9%)	
Alcohol						
Yes	115	8 (7,0%)	0,005*	46	10 (21,7%)	0,004*
No	28	7 (25,0)		82	13 (15,9%)	

Data are presented as headcount and percentage. P-value calculated from the ordered analysis of variances. (*) = significant P value < 0.05

➤ **Factors involved in the variation in cytopathological profiles between Niété and Yaoundé I**

Table 7 shows the variation in cervical cytological changes between the two study areas as a function of socio-demographic factors. A significant difference was found for ASCUS (p = 0.027) and LSIL (p = 0.017) lesions for the age factor. The prevalence of ASCUS type was higher in Yaoundé, irrespective of age group. The trend was reversed for the LSIL type where its prevalence was higher in Niété whatever the age group. For the marital status factor, a significant difference was found for LSIL lesions (p = 0.025). In fact, the prevalence of LSIL was higher among single women in Niété than in Yaoundé I. However, no significant difference between these profiles was observed for the study level factor.

Table 7: Factors socio-demographic factors involved in the variation in cytopathological profiles between Niété and Yaoundé I

Factors Socio-demographics	Cytological changes							
	ASCUS		AGUS		LSIL		HSIL	
Age	NIETE (%)	YDE I (%)	NIETE (%)	YDE I (%)	NIETE (%)	YDE I (%)	NIETE (%)	YDE I (%)
17-29	2,2	12,2	4,3	2,0	4,1	2,0	0,0	2,0
30-39	1,9	6,4	0,0	4,3	3,8	2,1	1,9	6,4
40-49	5,6	10,0	0,0	0,0	5,6	0,0	0,0	2,5
50-66	0,0	11,8	0,0	0,0	5,3	0,0	5,3	0,0
P	0,027*		0,136		0,017*		0,344	
marital status								
single	2,6	17,2	5,1	3,4	10,3	0,0	0,0	3,4
divorced	0,0	0,0	0,0	0,0	0,0	0,0	0,0	0,0
bride	2,7	6,5	0,0	0,9	0,0	0,0	0,9	0,0
widow	0,0	23,1	0,0	7,7	0,0	0,0	33,3	0,0
P	0,065		0,223		0,025*		0,223	
level of education								
No	0,0	5,9	8,3	0,0	0,0	0,0	0,0	0,0
primary	4,1	6,3	0,0	0,0	5,4	6,3	0,0	3,1
secondary	1,5	7,0	1,5	3,5	4,6	0,0	1,5	5,3
superior	0,0	10,8	0,0	2,7	0,0	0,0	50,0	0,0
P	0,141		0,329		0,155		0,269	

YDE= Yaoundé P-value calculated from ordered analysis of variances. (*) = significant P-value (< 0.05)

Table 8 shows the variation in cervical cytological changes between the two study areas according to the gynaecological-obstetric and toxicological factors of the women surveyed. A significant difference was found for AGUS lesions (p = 0.03) for the parity factor. For the number of sexual partners factor, a significant difference was found for LSIL-type lesions (p = 0.023). In fact, the prevalence of LSIL type was higher in Niété among women who had had more than three sexual partners, but the trend was reversed among those who had had only one partner. For the alcohol factor, a significant difference was found for ASCUS lesions (p = 0.027), the prevalence of ASCUS being higher in Yaoundé among women who consumed alcohol. A significant difference in profiles was observed in favour of Niété (7.1% versus 6.3%) in the HSIL type profile. However, no significant difference between these profiles was observed for age at first sexual intercourse. Among the factors associated with the different cytological changes in the women in our study, the age of the participants and alcohol consumption were responsible for the variation in ASCUS type cytopathological profiles between Niété and Yaoundé I. Parity was responsible for the variation in AGUS type. Parity was responsible for the variation in AGUS type, the number of sexual partners, marital status and again the age of the participants were responsible for the variation in LSIL type, and contraception, the variation in HSIL type.

Table 8: Gynaecological-obstetric and toxicological factors involved in the variation in cytopathological profiles between Niété and Yaoundé I

Gynaecological and obstetric factors toxicological	Cytological alterations							
	ASCUS		AGUS		LSIL		HSIL	
Parity	NIETE (%)	YDE I (%)	NIETE (%)	YDE I (%)	NIETE (%)	YDE I (%)	NIETE (%)	YDE I (%)
nulliparous	0,0	8,3	18,2	0,0	0,0	0,0	0,0	0,0
Paucipare	3,0	9,4	0,0	0,0	9,1	0,0	3,0	6,3
Multipare	2,8	10,0	0,0	2,7	3,7	1,8	0,9	2,7
P	0,313		0,03*		0,155		0,293	
APRS								
<15	0,0	8,0	0,0	0,0	4,0	4,0	4,0	8,0
15-19	3,5	8,3	1,7	2,1	3,5	1,0	0,9	3,1
20-24	0,0	17,9	0,0	3,6	16,7	0,0	0,0	0,0
P	0,228		0,367		0,152		0,292	
NPS								
1	2,1	6,8	0,0	2,3	2,1	2,3	0,0	3,4
2	0,0	7,1	0,0	3,6	0,0	0,0	0,0	3,6
≥3	3,4	18,9	2,3	0,0	6,8	0,0	2,3	2,7
P	0,106		0,082		0,023*		0,154	
Alcohol								
Yes	3,5	10,9	1,7	1,8	1,7	1,8	0,0	5,4
No	0,0	10,5	0,0	2,1	17,9	1,1	7,1	0,0
P	0,018*		0,136		0,333		0,333	
Contraception								
Yes	11,3	2,6	3,8	0,9	2,5	3,5	6,3	0,0
No	7,2	3,6	0,0	3,6	0,0	10,7	0,0	7,1
P	0,093		0,083		0,155		0,008*	

APRS = age of first sexual intercourse   = number of sexual partners. P-value calculated from the ordered analysis of variances. (*) = significant P value < 0.05

III.5 DISCUSSION

Early detection of cervical abnormalities is a useful and cost-effective tool for preventing pre-invasive lesions, particularly in resource-limited countries. The aim of this study was to carry out a comparative analysis of the cytopathological profiles of cervical cells from women in the Niété and Yaoundé I districts. In Niété, more than three quarters of the women enrolled in this study had macroscopic cervical abnormalities marked by inflammation. This could be explained by the existence of microbial infections, harmful chemical agents and poor hygiene. In this study, microbial causes were not investigated, but nevertheless, some studies have shown that yeast germs such as Candida albicans, (bacterial germs) and viruses such as HPV and herpes simplex virus type 2 were associated with these abnormalities (Koanga et al., 2014). Also, more than a third of these women used contraceptive methods. These physical and chemical stresses induced by this practice could probably explain some of the macroscopic abnormalities founded in this study. Our result corroborates with that found by Chiah in Algeria (2014), who had 98.74% inflammatory smears. Cervical cytological abnormalities, represented 9.8% in these women. Our results are similar to those obtained by Sankaranarayanan et al, in India (1998) using the VIA method, and by Koanga et al, in Douala (2014) who had 9.8% and 10.5% respectively. The high frequency of LSIL obtained in Niété was also observed in some studies, Chiah (2014) and Moukassa et al. in Congo-Brazzaville (2013) and Koanga et al. in Far North Cameroon (2016) who had 45.2% and 6.19% and 5.5% respectively. This could be explained by a lifestyle that exposes them (high rate for a number of sexual partners greater than three and for smoking). In addition, all the women surveyed in this area said that they had never had a cervical examination prior to our study, or that they were aware of cervical cancer or HPV. We therefore understand that these women presented to hospital on average at an advanced stage of these cervical cytological alterations. Furthermore, the frequency of LSIL in our study (2.6%) is comparable to the frequency obtained in the study by Mbakop et al. in Yaoundé (1996) in the case of HIV-positive women by cytological method (1.5%). The frequency of HSIL (1.3%) is comparable to that found in the Robbyr studies in Bafang- Cameroon (2002) and Somé et al. in Senegal (2016) who found 1.6%, one by the liquid method and the other by FCV. The frequency of ASCUS (4.6%) is similar to the work done by Koanga et al. in Far North Cameroon (2016), who found 3.8% using the cytological method.In Yaoundé I, more than three quarters of women had a normal cervix. This could be explained by appropriate hygiene, which limits the spread of harmful micro-organisms to the cervix. In addition, visits to a gynaecologist were very common. Cytological cervical abnormalities accounted for 16.3% of these women. Our results are similar to those obtained by Akinola et al. in Nigeria, who found 16.2% using the VIA method (2007). The high frequency of ASCUS obtained was also observed by the study done by Somé et al, in Senegal (2016) which had 2.5%. This could be explained by the physical and chemical stresses induced by the massive use of contraceptive methods by these women, and by alcohol consumption. In addition, the frequencies of HSIL (2%), AGUS (3.3%) and ASCUS (1.3%) found in this area are comparable to those obtained by ROBYR, (2002), Chiah in Algeria (2014) and Somé et al. (2016) who found respectively 2.5% by the liquid method; 3.91% and 2.5% by FCV.Comparative analysis of the profiles of the two zones showed a significant difference between the cytopathological profiles of the two zones. This can be explained by marked differences in mentalities, lifestyles and health problems specific to each zone. Indeed, despite the fact that a number of obstacles to access to care are generally recognised for rural communities, including geographical distance, the risk associated with transport, poor financial resources and lack of medical staff, it appears that, depending on country, age and lifestyle, geographical patterns of cervical cancer incidence or mortality vary greatly, sometimes to the advantage of rural communities and sometimes to their detriment (INSPQ, 2004). A comparison of the two pathological profiles showed that women in Yaoundé I were more prone to cervical cytological changes than those in rural areas (16.3% versus 9.8%). The same observation was made by Doll in England (1991), who found that among 13 populations studied (in as many different industrialised countries), incidence and mortality tended to be higher in urban areas for 23 of the 26 cancer sites considered. The study by Dangou et al. in Guinea-Conakry (2012) also revealed the same observation, with the urban environment (Conakry) recording more cases, including: 10% LSIL; 8% HSIL and 15.5% invasive cancer, compared with 5.2% LSIL; 15.3% HSIL and 10.5% invasive cancer in rural areas (Khorira). In contrast, the study by Moukassa et al. in Congo-Brazzaville (2013) found that among women who consented to this study, the relative frequency of cervical intraepithelial neoplasia (EIN) was higher in rural areas than in urban areas (14.3% versus 7.1%, $p \leq 0.05$). Determining the factors involved in the variation in cytopathological profiles between the two areas enabled us to understand why the urban area presented more cases than the rural area. The factors found to be responsible for this variation were: the age of the participants, their marital status, the number of sexual partners, parity, alcohol consumption and contraception. More women in urban areas (43.4% versus 29.3%) presented with cytological changes to the cervix from the age of 30. Multiparous women, whose exposure to trauma increases with the number of births they have had, were also more likely to be found in urban areas (17.3% versus 7.3%). They also consumed more alcohol (21.7% versus 7%), and had more sexual partners with more than three (21.6% versus 14.8%). Women in urban areas were therefore more likely to have higher pathological smear tests than those in rural areas, because the women present were of a higher risk age ($\geq$30 years), with a higher number of partners over three, as were alcohol consumption and the number of sexual partners.

CONCLUSION RECOMMENDATIONS AND OUTLOOK

CONCLUSION

At the end of our study, which involved a comparative analysis of the cytopathological profiles of the cervical cells of patients living in the districts of Niété and Yaoundé I, we found that 9,8% of smears were pathological in Niété, with 4.6% ASCUS, 1.3% AGUS, 2.6% LSIL, 1.3% HSIL, and 16.3% of smears were pathological in Yaoundé I, with 1.3% ASCUS, 3.3% AGUS, 9.8% LSIL and 2% HSIL. In addition, we observed a significant difference between the profiles of these two study settings (P= 0.034; χ^2 = 11.01), the urban setting being more prone to cervical cytological alterations than the rural setting (Niété). Age, parity and marital status are factors explaining this high rate. In addition, the age of the participants and alcohol consumption were found to be responsible for the variation in ASCUS type cytopathological profiles, parity was responsible for the variation in AGUS type, the number of sexual partners, marital status and again the age of the participants were responsible for the variation in LSIL type, and contraception, the variation in HSIL type. There is therefore an urgent need to organise well-targeted cervical cancer screening programmes in Cameroon, using a methodology that is acceptable in terms of cost and effectiveness. The population is concerned about cancer, and education campaigns can help raise awareness.

RECOMMENDATION

Cervical cancer remains a major public health problem, the second most common cause of cancer in women after breast cancer. Despite the considerable success of cytological screening in preventing cervical cancer, the smear test has not been as successful as might be expected in reducing its incidence on a large scale. What's more, when it comes to preventing cervical cancer, screening seems to benefit only a tiny proportion of the world's population, while a large proportion of those who do benefit endure its weaknesses. We therefore recommend that the Cameroonian government organise large-scale cervical cancer screening campaigns in both rural and urban areas, step up care for women living with precancerous or cancerous lesions, and take into account new data on the epidemiology o f cervical cancer in Cameroon.

This study revealed that the quality of cytological screening, as performed, was unsatisfactory. One of the factors blamed was the need for the cytopathologist to read a high number of smears per day (30-40 smears/day) because of the constraints of the study, especially in rural areas. In current practice, apart from screening campaigns, if a smear is taken in Niété, the result does not reach the patient until three weeks later, as it has to be sent to the capital, Yaoundé. This means that patients are lost in the follow-up process, even if surgical options are available. What's more, in rural areas of Cameroon, there are no means of carrying out a colposcopic examination. Another factor was the high proportion of inflammation (cervicitis) found in the Niété population, which sometimes made it difficult to read the results.

OUTLOOK

In view of the various results obtained in this study, the next step will be to extend the study to several other localities in Cameroon, so as to be able to generalise the results throughout the country, and to use the molecular aspect to explain the causes of the variation in profiles between urban and rural areas. The aim is to considerably reduce the prevalence of cervical cancer in Cameroon, and to open up new avenues for the prevention and management of people living with precancerous and cancerous lesions of the cervix.

ACCP, 2004. Preventing Cervical Cancer Worldwide - PRB-ACCP_PreventCervCancer_EN.pdf [WWW Document]. URL http://screening.iarc.fr/doc/PRB-ACCP_PreventCervCancer_FR.pdf (accessed 10.24.16).

Akinola, O.I., Fabamwo, A.O., Oshodi, Y.A., Banjo, A.A., Odusanya, O., Gbadegesin, A., Tayo, A., 2007. Efficacy of visual inspection of the cervix using acetic acid in cervical cancer screening: A comparison with cervical cytology. J. Obstet. Gynaecol. 27, 703- 705. doi:10.1080/01443610701614421

Baba, A.I., Câtoi, C., 2007. Comparative Oncology. The Publishing House of the Romanian Academy, Bucharest.

Benchimol, 2014. Dysplasia of the uterine cervix [WWW Document]. URL https://docteur- benchimol.com/gynecology/18-cervical-dysplasia.html (accessed 10.26.16).

Boman, F., Duhamel, A., Trinh, Q.D., Deken, V., Leroy, J.-L., Beuscart, R., 2003. Evaluation of cytological screening for cervical cancer and precancerous lesions. Bull. Cancer (Paris) 90, 643-647.

Bosch, F., Broker, T., Forman, D., Moscicki, A., Gillison, M., Doorbar, J., Stern, P., Stanley, M., Arbyn, M., Poljak, M., Cuzick, J., Castle, P., Schiller, J., Markowitz, L., Fisher, W., Canfell, K., Denny, L., Franco, E., Steben, M., Kane, M., Schiffman, M., Meijer, C., Sankaranarayanan, R., Castellsagué, X., Kim, J., Brotons, M., Ale-many, L., Albero, G., Diaz, M., de Sanjosé, S., 2013. Comprehensive control of human papillomavirus infections and related diseases. - PubMed - NCBI [WWW Document]. URL https://www.ncbi.nlm.nih.gov/pubmed/24229716 (accessed 10.24.16).

Bray, F., Ren, J.-S., Masuyer, E., Ferlay, J., 2013. Global estimates of cancer prevalence for 27 sites in the adult population in 2008. Int. J. Cancer 132, 1133-1145. doi:10.1002/ijc.27711

Carozzi, F.M., Ocello, C., Burroni, E., Faust, H., Zappa, M., Paci, E., Iossa, A., Bonanni, P., Confortini, M., Sani, C., 2016. Effectiveness of HPV vaccination in women reaching screening age in Italy. J. Clin. Virol. Off. Publ. Pan Am. Soc. Clin. Virol. 84, 74-81. doi:10.1016/j.jcv.2016.09.011

Chard, 1994. The Uterus by Chard, T. (EDT)/ Grudzinskas, J. G. (EDT): Cambridge University Press 1994-11-17, Cambridge 9780521424530 paperback Blackwell'shttp://www.abebooks.com/servlet/BookDetailsPL?bi=19472347867&

searchurl=isbn% 3D0521424534%26sortby%3D17 (accessed 10.24.16).

Chiah, B., 2014. Contribution à l'étude du dépistage du cancer du cerv de l'utérus au niveau de la wilaya de Bechar et la recherche du Papillomavirus humain par la réaction de polymérisation en chaine [WWW Document]. URL http://dspace.univ- tlemcen.dz/bitstream/112/5572/1/MEMOIRE.pdf (accessed 1.5.17).

CNGOF, 2016. La colposcopie | Exploration du cerv uterin [WWW Document]. URL http://www.cngof.fr/interventions-gynecologiques/352-la-colposcopie-exploration-du- col-uterin (accessed 10.29.16).

Criton, C., 2010. Precancerous lesions of the uterine cervix and HIV infection - SidaSciences [WWW Document]. URL http://sidasciences.inist.fr/?Lesions-precancereuses-du-col (accessed 10.24.16).

Dallenbach, H.G., 2006. Color Atlas of Histopathology of the Cervix Uteri. Springer Berlin Heidelberg, Berlin, Heidelberg.

Dangou,jean-marie, 2012. Prevention and management of cervical cancer in Guinea https://www.aho.afro.who.int/fr/ahm/issue/15/reports/pr%C3%A9vention-et-management-du-cancer-du-col-ut%C3%A9rin-en-guin%C3%A9e (accessed 1.17.17).

DIOURI MOHAMED KHALIL, 2008. CERVICAL CANCER SCREENING IN THE PREFECTURES OF RABAT AND SKHIRAT TEMARA: CURRENT SITUATION AND PERSPECTIVES [WWW Document].URL http://fulltext.bdsp.ehesp.fr/Inas/Memoires/massp/sp/2008/7508.pdf(accessed 10.24.16).

Doll, R., 1991. Urban and rural factors in the aetiology of cancer. Int. J. Cancer 47, 803-810.

Embolo, E., Martin, K.M., Albert, M.S., Annie, N.N., 2016. Prevalence of precancerous lesions of the uterine cervix according to VIA, VILI and cytological aspect: analysis of usefulness of combination. Int. J. Res. Biosci. Volume 5 Issue 1.

F. Lecuru, 2008. GM001titres2004.qxd - 2008_GO_313_lecuru.pdf [WWW Document]. URL http://www.cngof.asso.fr/d_livres/2008_GO_313_lecuru.pdf (accessed 10.25.16).

Faysal Ali Saksouk, 2016. Cervical Cancer Imaging: Overview, Nuclear Imaging, Computed Tomography (CT).

G. Ronco, 2009. Cervical Cancer Screening in the European Union [WWW Document]. URL http://zora.onko-

i.si/fileadmin/user_upload/dokumenti/novice/EJC_Cervical_screening_in_EU_20
09.p df (accessed 10.29.16).

Golijow, C.D., Abba, M.C., Mourón, S.A., Laguens, R.M., Dulout, F.N., Smith,
J.S., 2005. Chlamydia trachomatis and Human papillomavirus infections in
cervical disease in Argentine women. Gynecol. Oncol. 96, 181-186.
doi:10.1016/j.ygyno.2004.09.037
Healthline, 2012. Pap smear [WWW Document]. URL
http://fr.healthline.com/health/test-de-papanicolaou#Pr%C3%A9paration4
(accessed 1.14.17).
IARC, 2013. Latest global cancer statistics - pr223_E.pdf [WWW Document].
URL https://www.iarc.fr/fr/media-centre/pr/2013/pdfs/pr223_F.pdf (accessed
10.24.16).
INC, 2013. Risk Factors - Cervical Cancer | Institut National Du Cancer [WWW
Document]. URL http://www.e-cancer.fr/Patients-et-proches/Les-
cancers/Cancer-du-col-de-l-uterus/Risk-factors (accessed 10.26.16).
INSPQ, 2004. Does living in a rural community really make a difference to your
health and well-being? | INSPQ - Institut national de santé publique du Québec
James, P., 2016. 1-s2.0-S2214854X1530008X-main.pdf - nott nabs cx anat rev
tria 16.pdf
http://eprints.whiterose.ac.uk/100807/1/nott%20nabs%20cx%20anat%20rev%20t
ria% 2016.pdf (accessed 10.24.16).
J.-C. Boulanger, 2010. Gynaecology - Presentation - EM consults [WWW
Document]. URL http://www.em-consulte.com/article/277306/colposcopie
(accessed 10.25.16).
Kemfang, J.D.N., Ngassam, A., Meka, E.N. um, Fouogue, J.T., Tagne, J.C.,
Sando, Z., Mendounga, J.T.N., Kasia, J.M., 2015. Screening for Cervical Cancer
by Visual Inspection of the Cervix after Application of Acetic Acid in Yaoundé,
Cameroon. Health Sci. Dis. 16.
Khenchouche, A., Sadouki, N., Boudriche, A., Houali, K., Graba, A., Ooka, T.,
Bouguermouh, A., 2013. Human Papillomavirus and Epstein-Barr virus co-
infection in Cervical Carcinoma in Algerian women. Virol. J. 10, 340.
doi:10.1186/1743-422X- 10-340
Koanga, M.L.M., Annie, N.N., Grace, N., Michel, W., Charlotte, B.E., Henri,
A.Z.P., 2014. Association of Cervical Inflammation and Cervical abnormalities
in Women Infected with Herpes Simplex virus 2 (HSV2). Int. J. Trop. Med.
Public Health 4, 10-14.
Koanga, M.M.L., 2016. Worrying and real risk for human papilloma virus-

induced cervix lesions in cameroonian women: insight into a socioeconomically disadvantaged Muslim region | International Journal Of Development Research [WWW Document]. URL

http://www.journalijdr.com/worrying-and-real-risk-human-papilloma-virus-induced- cervix-lesions-cameroonian-women-insight-socioe (accessed 1.12.17).

Koss, L.G., 1989. The Papanicolaou test for cervical cancer detection. A triumph and a tragedy. JAMA 261, 737-743.

Koutsky, L.A., Holmes, K.K., Critchlow, C.W., Stevens, C.E., Paavonen, J., Beckmann, A.M., DeRouen, T.A., Galloway, D.A., Vernon, D., Kiviat, N.B., 1992. A cohort study of the risk of cervical intraepithelial neoplasia grade 2 or 3 in relation to papillomavirus infection. N. Engl. J. Med. 327, 1272-1278. doi:10.1056/NEJM199210293271804

Leidy, N.K., 1999. Cost-effectiveness of Methods to Enhance Sensitivity of Papanicolaou Testing. JAMA 282, 1419. doi:10.1001/jama.282.15.1419

Letian, T., Tianyu, Z., 2010. Cellular receptor binding and entry of human papillomavirus.

Virol. J. 7, 2. doi:10.1186/1743-422X-7-2

Luna, J., Plata, M., Gonzalez, M., Correa, A., Maldonado, I., Nossa, C., Radley, D., Vuocolo, S., Haupt, R.M., Saah, A., 2013. Long-term follow-up observation of the safety, immunogenicity, and effectiveness of Gardasil™ in adult women. PloS One 8, e83431. doi:10.1371/journal.pone.0083431

Maclean, D., 2009. HPV, Cervical Dysplasia and Cervical Cancer | CATIE - Canada's Source for HIV and Hepatitis C Information

Mantovani, F., Banks, L., 2001. The human papillomavirus E6 protein and its contribution to malignant progression. Oncogene 20, 7874-7887. doi:10.1038/sj.onc.1204869

Mbakop, A., Zekeng, L., Mbassi, J.R., Essimbi, F., 1996. [Cytologic aspects of cervical smears in optic microscopy in HIV seropositive women in Yaounde-Cameroon (Central Africa)]. Arch. Anat. Cytol. Pathol. 44, 250-253.

McCredie, M.R.E., Sharples, K.J., Paul, C., Baranyai, J., Medley, G., Jones, R.W., Skegg, D.C.G., 2008. Natural history of cervical neoplasia and risk of invasive cancer in women with cervical intraepithelial neoplasia 3: a retrospective cohort study. Lancet Oncol. 9, 425-434. doi:10.1016/S1470-2045(08)70103-7

Miller, A.B., 2012. Advances in Cancer Screening. Springer Science & Business Media.

Monsonégo, J., 2006. Preventing cervical cancer: challenges and prospects for HPV vaccination. Gynécologie Obstétrique Fertil. 34, 189-201. doi:10.1016/j.gyobfe.2006.01.036

Moscicki, A.-B., Schiffman, M., Burchell, A., Albero, G., Giuliano, A., Goodman, M.T., Kjaer, S. K., Palefsky, J ., 2012. Updating the Natural History of Human Papillomavirus and Anogenital Cancers. Vaccine 30, F24-F33. doi:10.1016/j.vaccine.2012.05.089

Mougin, C., Dalstein, V., Prétet, J.L., Gay, C., Schaal, J.P., Riethmuller, D., 2001. [Epidemiology of cervical papillomavirus infections. Recent knowledge]. Presse Medicale Paris Fr. 1983 30, 1017-1023.

Moukassa, D., 2013. Comparative study of risk factors for precancerous cervical lesions in two Congolese health departments (Congo-Brazzaville).

Moukassa, D., N'Golet, A., Lingouala, L.G., Eouani, M.L., Samba, J.B., Mambou, J.V., Ompaligoli, S., Moukengue, L.F., Taty-Pambou, E., 2007. [Precancerous lesions of the uterine cervix in Pointe-Noire, Congo]. Med. Trop. Rev. Corps Sante Colon. 67, 57-60.

Muñoz, N., Castellsagué, X., de González, A.B., Gissmann, L., 2006. Chapter 1: HPV in the etiology of human cancer. Vaccine 24 Suppl 3, S3/1-10. doi:10.1016/j.vaccine.2006.05.115

Nkegoum, B., Belley Priso, E., Mbakop, A., Gwent Bell, E., 2001. [Precancerous lesions of the uterine cervix in Cameroonian women. Cytological and epidemiological aspects of 946 cases]. Gynecol. Obstet. Fertil. 29, 15-20.

Norstrom A., T. Radberg, 2002. The problem of cervical cancer screening. http://www.canceraquitaine.org/sites/default/files/documents/INFOS-PRO/referentiels/gynecologie-senologie/Ref-Col-0905.pdf (accessed 10.24.16).

WHO,2014.Cancer Country Profile - cmr_en.pdf [WWW Document]. URL http://www.who.int/cancer/country-profiles/cmr_fr.pdf?ua=1 (accessed 10.19.16).

WHO, 2007a. Cervical cancer control. Essential practices guide [WWW Document]. URL http://docplayer.fr/5541941-La-lutte-contre-le-cancer-du-cervix-uterus-essential-practices-guide.html (accessed 1.14.17).

WHO, 2007b. Cervical cancer control English-21.indd - text_en.pdf [WWW Document]. URL http://screening.iarc.fr/doc/text_fr.pdf (accessed 10.24.16).

OUEDRAOGOTeega,2015.labiogene_m2_ouedraogo_teega-wende_clarisse.pdf.pdf http://www.labiogene.org/IMG/pdf/labiogene_m2_ouedraogo_teega-wende_clarisse.pdf.pdf (accessed 10.26.16).

P. Lopes, 2013. Viral-induced pelvic cancers: cervical cancer [WWW Document]. URL http://www.lesjta.com/article.php?ar_id=1547 (accessed 10.25.16).

P.M. TEBEU, 2005. Precancerous lesions of the uterine cervix in rural areas: cross-sectional study - 5201_6.pdf
http://www.gfmer.ch/Presentations_Fr/Pdf/5201_6.pdf (accessed 10.24.16).
Potischman, N., Brinton, L.A., 1996. Nutrition and cervical neoplasia. Cancer Causes Control CCC 7, 113-126.
Renolleau, C., 1996. Practical conduct in gynaecology. Heures de France.

ROBYR, R., 2002. Pilot study of cervical cancer screening in a rural area of Cameroon Cameroon [WWW Document].URL
http://www.unige.ch/cyberdocuments/theses2002/RobyrR/these.html (accessed 10.24.16).
Sando, Z., Fouogue, J.T., Fouelifack, F.Y., Fouedjio, J.H., Mboudou, E.T., Oyono, J.L., 2014. Profile of gynaecological and breast cancers in Yaoundé - Cameroon. Pan Afr. Med. J. 17. doi:10.11604/pamj.2014.17.28.3447
Sankaranarayanan, R., Wesley, R., Somanathan, T., Dhakad, N., Shyamalakumary, B., Amma, N.S., Parkin, D.M., Nair, M.K., 1998. Visual inspection of the uterine cervix after the application of acetic acid in the detection of cervical carcinoma and its precursors. Cancer 83, 2150-2156.
Sawaya, G.F., McConnell, K.J., Kulasingam, S.L., Lawson, H.W., Kerlikowske, K., Melnikow, J., Lee, N.C., Gildengorin, G., Myers, E.R., Washington, A.E., 2003. Risk of cervical cancer associated with extending the interval between cervical-cancer screenings. N. Engl. J. Med. 349, 1501-1509.
doi:10.1056/NEJMoa035419
Schiffman, M.H., Brinton, L.A., 1995. The epidemiology of cervical carcinogenesis. Cancer 76, 1888-1901.
Sellors, J.W., 2004. Colposcopy and Treatment of Cervical Intraepithelial Neoplasia: A Beginner's Manual [WWW Document]. URL
http://screening.iarc.fr/colpo.php?lang=2 (accessed 1.14.17).
Siegel, R.L., Miller, K.D., Jemal, A., 2016. Cancer statistics, 2016. CA. Cancer J. Clin. 66, 7-30. doi:10.3322/caac.21332

Smith, J.S., Herrero, R., Bosetti, C., Muñoz, N., Bosch, F.X., Eluf-Neto, J., Castellsagué, X., Meijer, C.J.L.M., Van den Brule, A.J.C., Franceschi, S., Ashley, R., International Agency for Research on Cancer (IARC) Multicentric Cervical Cancer Study Group, 2002. Herpes simplex virus-2 as a human papillomavirus cofactor in the etiology of invasive cervical cancer. J. Natl. Cancer Inst. 94, 1604-1613.

Solomon, D., Davey, D., Kurman, R., Moriarty, A., O'Connor, D., Prey, M., Raab, S., Sherman, M., Wilbur, D., Wright, T., Young, N., Forum Group Members, Bethesda 2001 Workshop, 2002. The 2001 Bethesda System: terminology for reporting results of cervical cytology. JAMA 287, 2114-2119. Somé, O.-R., Zongo, N., Ka, S., Wardini, R., Dem, A., 2016. Mass cervicovaginal smear screening: results of an African experience. Gynecology Obstetrics Fertil. 44, 336-340. doi:10.1016/j.gyobfe.2016.04.006 Soumaya OUCHEN, 2009.THE PLACE OF THE CYSTOSCOPY IN THE CLASSIFICATION OF UTERINE COLUMN CANCER [WWW Document]. URLhttp://wd.fmpm.uca.ma/biblio/theses/annee-htm/FT/2009/these54-09.pdf (accessed 10.26.16).
Thin, R.N., Atia, W., Parker, J.D., Nicol, C.S., Canti, G., 1975. Value of Papanicolaou-stained smears in the diagnosis of trichomoniasis, candidiasis, and cervical herpes simplex virus infection in women. Br. J. Vener. Dis. 51, 116-118.
Trimble, C.L., Clark, R.A., Thoburn, C., Hanson, N.C., Tassello, J., Frosina, D., Kos, F., Teague, J., Jiang, Y., Barat, N.C., Jungbluth, A.A., 2010. Human papillomavirus 16- associated cervical intraepithelial neoplasia in humans excludes CD8 T cells from dysplastic epithelium. J. Immunol. Baltim. Md 1950 185, 7107-7114. doi:10.4049/jimmunol.1002756
WOLFGANG, K., 2003. Color Atlas of Cytology, Histology and Microscopic Anatomy [WWW Document]. URL http://www.thieme.com/books-main/books/product/3386- color-atlas-of-cytology-histology-and-microscopic-anatomy (accessed 1.14.17).

APPENDICES

Appendix 1: Papanicolaou staining

The first step consists of gradually rehydrating the samples by successively passing the slides through :

- two vats of alcohol, one at 95° and the other at 80⁰ for 30s each;
- a vat of 70° alcohol for 30 seconds;
- a vat of 50° alcohol for 30 seconds;
- two tanks of distilled water for 10s each.

The slides were then stained in the first reagent, Harris Haematoxylin, for 3 minutes before being washed in 3 successive tanks of distilled water for 10 seconds each. Progressive dehydration was carried out, followed by a second staining in G6 orange:

- a vat of 50° alcohol for 30 seconds;
- a vat of 70° alcohol for 30 seconds;
- a vat of 80° alcohol for 30 seconds;
- a vat of 95° alcohol for 30 seconds;
- a tray of G6 orange for 15s.

A double dehydration is then carried out to achieve colouration by the third reagent:

- two vats of 95° alcohol for 15 seconds each;
- a vat of polychrome EA50 for 1 minute;

The final stage gradually led to the assembly of the blades:

- two vats of 95° alcohol for 30 seconds;
- a vat of absolute alcohol for 5 minutes;
- a vat of the absolute alcohol-xylene mixture for 10s;
- a xylene mounting tray.

The mounting step consisted of placing a few drops of Eukitt on the slide and adhering the coverslip to it without leaving any air bubbles.

Appendix 2: Interpretation of slides according to the Bethesda system.

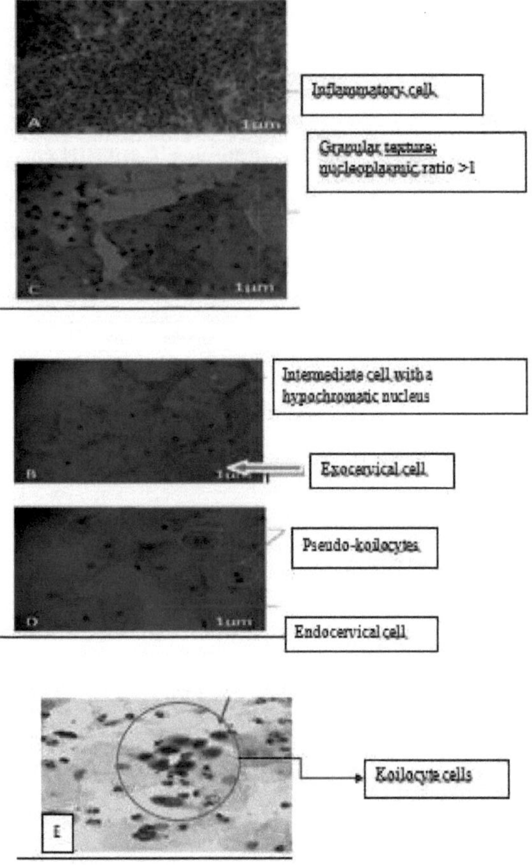

A: *inflammatory cells;*

B: *ASCUS lesion (cells with a hypochromatic nucleus);*

C: *AGUS lesion (in basal cells with a high nucleus/cell ratio and granular texture);*

D: *Intraepithelial lesions of low-grade squamous cells, LSIL (koilocyte-like cells);*

E: *High-grade squamous intraepithelial lesions, HSIL (koilocytic cells)*

Appendix 3: Ethical Clearance

REPUBLIQUE DU CAMEROUN Paix - Travail- Patrie UNIVERSITE DE DOUALA		REPUBLIC OF CAMEROON Peace - Work - Fatherland UNIVERSITY OF DOUALA

COMITE D'ETHIQUE INSTITUTIONNEL
DE LA RECHERCHE
POUR LA SANTE HUMAINE

Arrêté N° 0977/Minsanté/SESP/SG/DROS du 16 avril 2012 portant création, organisation et fonctionnement des comités d'éthique de la recherche pour la santé humaine au sein des structures relevant du Ministère en charge de la santé publique

N° CEI-UDo/707/ 11/ 2016/T   Douala, le 24 Novembre 2016

CLAIRANCE ÉTHIQUE

Le Comité d'Ethique Institutionnel de la Recherche pour la Santé Humaine de l'Université de Douala (CEI-UDo) en sa session du 24 Novembre 2016, a examiné le projet de recherche intitulé «**Analyse comparative du profil cytopathologique des cellules cervicales des patientes résidentes dans les arrondissements de Niété et de Yaoundé I**» soumis par **EBODE BALLA FOUDA Engelbert,** tenant lieu de Mémoire à la Faculté des Sciences de l'Université de Douala.

Le présent projet de recherche est d'un intérêt scientifique certain et ne présente aucun risque pour le participant. Les objectifs et la méthodologie de l'étude sont clairement décrits. Le formulaire de consentement éclairé est bien élaboré. Le principe de confidentialité des données est respecté. Les compétences requises pour la supervision des travaux de recherche sont présentes.

Au vu de ce qui précède, le CEI-UDo approuve pour une durée d'un an, la mise en œuvre de la présente version du protocole.

EBODE BALLA FOUDA Engelbert est responsable du respect scrupuleux du protocole et ne devrait y apporter aucun amendement aussi mineur soit-il, sans avis favorable du CEI-UDo. Les investigateurs sont tenus de collaborer avec le CEI-UDo pour le suivi des aspects éthiques du protocole approuvé. Le rapport final du projet de recherche devra être déposé au CEI-UDo pour archivage.

La présente clairance éthique est délivrée pour servir et valoir ce que de droit. Elle peut être annulée en cas de non-respect de la réglementation en vigueur et des recommandations sus-mentionnées.

<u>Ampliations</u>
  - MINSANTE

LE PRESIDENT

Pr Léopold Gustave LEHMAN

NB : Il n'est délivré qu'un seul exemplaire de la clairance éthique.

Appendix 4: Data collection form

<table>
<tr><td rowspan="3">UNIVERSITE DE DOUALA</td><td colspan="2">Thème :</td></tr>
<tr><td>FORMULAIRE DE COLLECTE DES DONNEES

Participation à une recherche biomédicale</td><td>N° Code Protocole : /___/___/___/

Anonymat : /___/___/___/

Date: /___/___/___ ___/

N° Téléphone :</td></tr>
<tr><td colspan="2">Analyse comparative du profil cytopathologique des cellules cervicales de patientes résidentes dans les Arrondissements de Niété et de Yaoundé</td></tr>
</table>

Informations Sociales et Anthropologiques (A)

(1)-Noms et Prénoms : /_ _ _ _ _ _ _ _ _ _ _
_ _ _ _ _ _ _ _ _ _ _ _ _ _ _ _/

(2)- Sexe : (Cocher)   M /___/   F /___/

(3)-Date et lieu de naissance : J/M/A
/___ ___/___ ___/___ ___ ___/___/___ ___ ___/

(4)-N° de tél : /__/__/__/__/__/__/__/__/__/

(5)-Quartier de résidence : /___ ___ ___ ___ ___ ___/

(6)-Statut Matrimonial : /___ ___ ___ ___ ___ ___/

(7)-Profession et Niveau D'étude: /___ ___ ___ ___/ ___ ___ ___ ___ ___/

(8)- Poids /___ ___ ___ ___ ___ ___/

(9)- Taille /___ ___ ___ ___ ___/

(10)- Ethnie /___ ___ ___ ___ ___ ___/

Données Gynécologiques et Obstétriques (B)

(1)- Date des Dernières Règles (DDR) : /___ ___/___ ___/___ ___ ___/ et saignements pendant les rapports

(2)-Age du Premier Rapport Sexuel (APRS) : /___ ___/

(3)- Nombre de Partenaires Sexuels (NPS) : /___ ___ ___/

(4)- Age du Premier Saignement (APS) : /___ ___ ___/

(5)- Methode Contraceptive : OUI /___/ NON /___/

(6)- Nombre de Grossesses : /___ ___ ___/

---

(7)-Age de la Première Grossesse : /___ ___/

(8)- Nombre d'accouchements à Terme : /___ /

(9)- Avez-vous des Troubles de Menstruation : /___/

Information sur la Pathologie et Statut Sérologique (C)

(1)-Avez Déjà Entendu Parler Du Frottis Cervico-Vaginal (FCV): /___/

(2)- Avez Déjà Entendu Parler Du cancer Du Col De l'Utérus /___/

(3)- Avez Déjà Entendu Parler Du Virus Des Papillomes Humains /___ ___/

(4)- Combien De Fois Voyez-Vous Un Gynécologue Par An : /___/

(5)- Quel est Votre Statut Sérologique VIH : /___ ___/

(6)- Quel est Votre Statut Sérologique Hépatite B et C : /___ ___/ et /___ ___/

Aspect toxicologique (D)

(1)- Consommez-vous Du Tabac ? /___ ___/

(2)- Avez-vous une personne dans votre Entourage Proche Qui Fume ? /___ ___/

(3)- Consommez- vous De l'Alcool ? /___ ___/

(4)- Consommez-vous d'Autres stupéfiants (café, thé) ? : /___/

56

 **MIX**
Papier aus verantwortungsvollen Quellen
Paper from responsible sources
FSC® C105338

Printed by Books on Demand GmbH, Norderstedt / Germany